37(02)

Natural
and human activity

A level

Geography
Studies
Arthur Byrne
Martin Thom

To Kathleen Byrne and Mary O'Brien with grateful thanks.
Also –
Love and thanks to the Clova dream team,
Ailsa, Cathy, Laura and Ellen

All rights reserved. No part of this publication may be reproduced, stored in a retrieval system or transmitted in any form or by any means, electronic, mechanical, photocopying, scanning, recording or otherwise, without the prior written permission of the copyright owners and publisher of this book.

6 5 4 3 2

© Arthur Byrne and Martin Thom
2002

Designed by Colourpoint Books.
Printed by Nicholson & Bass Ltd.

Cover: Lightning on Sugar Loaf Mountain, Brazil.
(Getty Images).

ISBN 1 898392 88 9

Colourpoint Books
Unit D5, Ards Business Centre
Jubilee Road
NEWTOWNARDS
County Down
Northern Ireland
BT23 4YH
Tel: 028 9182 0505
Fax: 028 9182 1900
E-mail: info@colourpoint.co.uk
Web-site: www.colourpoint.co.uk

Arthur Byrne, MSc, BSSc (Hons), PGCE, PGC in Marketing, is Head of Geography at St Louis Grammar, Ballymena. He graduated from Queen's University Belfast, in 1975 and completed his Post-Grad in Education at Warwick University. He taught for two years at Cardinal Wisemann Comprehensive in Coventry, before his appointment to St Louis Grammar. He was appointed Head of Geography in 1982, and has had wide experience in teaching and assessing geography at A Level.

Martin Thom BA (QUB 1981), PGCE (Stranmillis College 1982), taught geography at secondary level in a number of schools in Bangor and Belfast before moving to Sullivan Upper School, Holywood, in 1989. He later became head of department when Jim Robinson retired. His interests include reading, travel and people.

Introduction

'All humanity lives on the fragile interface between Earth and sky.' For many people this is merely a fact but for others it can be a terrifying reality. While the sky above us and the Earth below provide vital resources for our survival and comfort, the same two realms are the sources of violence and devastation. Across the face of this planet extreme meteorological events such as hurricanes, monsoons, droughts and storms or tectonic activity in the form of earthquakes or volcanic eruption, can decimate the landscape, people and property.

Such episodes with their threat to life we term **natural hazards**.

Natural hazards are as old as the Earth itself. Some writers define them as environmental hazards, which includes both natural and human events.

In this book a natural hazard is defined as an extreme natural event or process that takes place in an area of human settlement and could cause loss of life and damage to the economic and environmental systems. Thus deforestation, desertification and rising sea levels are regarded as environmental problems rather than natural hazards. Nagle (1998) suggested aspects such as their magnitude, frequency, duration, areal extent, spatial concentration and speed of onset could further define natural hazards. This book examines these hazards, their impact and the response of people both to the menace and their occurrence. It is designed to provide a resource for CCEA's A-level geography Module 4 Unit D, 'Natural Hazards and Human Activity'.

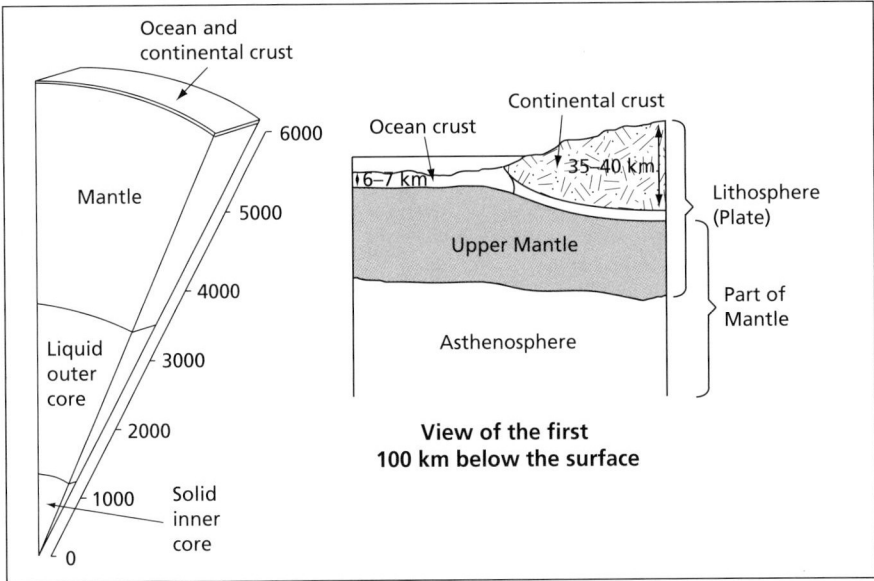

Resource 1: The Earth's internal structure

Chapter 1

'Global Revolution'

Historically sudden and dramatic incidents such as volcanic eruptions and earthquakes have a key role in people's worldview and mythology. Such seemingly random events were given the guise of divine wrath or justice, but while human awareness of these incidents is ancient, the explanation of them is modern and we have developed the literally ground breaking concepts of **plate tectonics**.

The theory of plate tectonics

The history of the development of this theory lies beyond the scope of this book; rather, we will examine the outcome of that process.

Volcanoes and earthquakes provide us with the essential evidence to build a picture of the Earth's interior. The deepest mines only scratch the Earth's surface layers but volcanoes bring up material from below that we can study and by examining also how the Earth shudders during earthquakes an image of our planet's structure has been created (**Resource 1**).

New research continues to modify this model including, in recent years, the shape of the core and, using meteorite analysis, its chemical nature. The section of the model that concerns us most here is the crust and the upper mantle. In 1915 Alfred Wegener, a German meteorologist, published a book stating his belief that the world's continents were not fixed in location but had separated and drifted apart from one ancient megacontinent he named Pangaea ('all land'). Only decades after his death in 1930 was conclusive evidence for his concept uncovered, though the mechanism he proposed for shifting continents has been proved invalid. The key which unlocked the puzzle of parallel shorelines and fossil distributions was the mapping of the deep ocean floor. Sonar depth sounding, developed in marine warfare during the second world war, was used to chart the seabed of the Atlantic in 1947. This identified in the depths of the ocean a long chain of underwater mountains, the Mid-Atlantic ridge running parallel to the Atlantic coastline and consisting of young volcanic rocks. Earthquake epicentres matched the line of this mountain chain and the deep central rift valleys that run down its length (**Resources 2 & 3**).

The Indian and Pacific Oceans contain similar ridges as well as deep ocean trenches close to, and parallel with, coastal mountains. The theory links these features as zones of construction and destruction of oceanic crustal material (**Resource 2**).

The model states that the lithosphere consists of up to 14 large sections or plates. Some of these form the ocean basins, some carry continents and some have both. By the 1970s scientists, using cameras carried by submersibles, had watched submarine volcanic eruptions along the mid-ocean ridges. This was plate construction or **sea floor spreading**. At the same time as new crust is created, the theory suggested, elsewhere crust must be being destroyed or the world would be getting larger. The distribution of many volcanoes and earthquakes parallel to deep-ocean trenches suggested sites for this destruction or **subduction** of plate material (**Resources 2 & 3**).

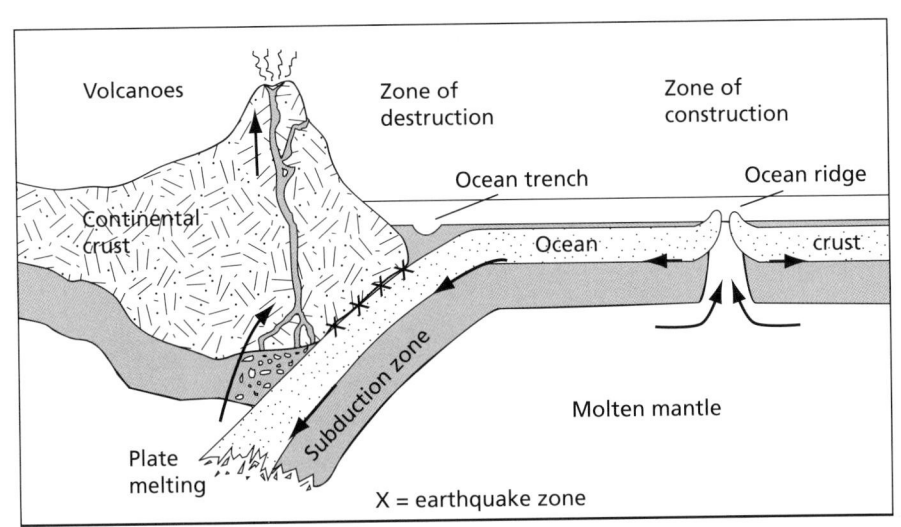

Resource 2: Plate construction and destruction

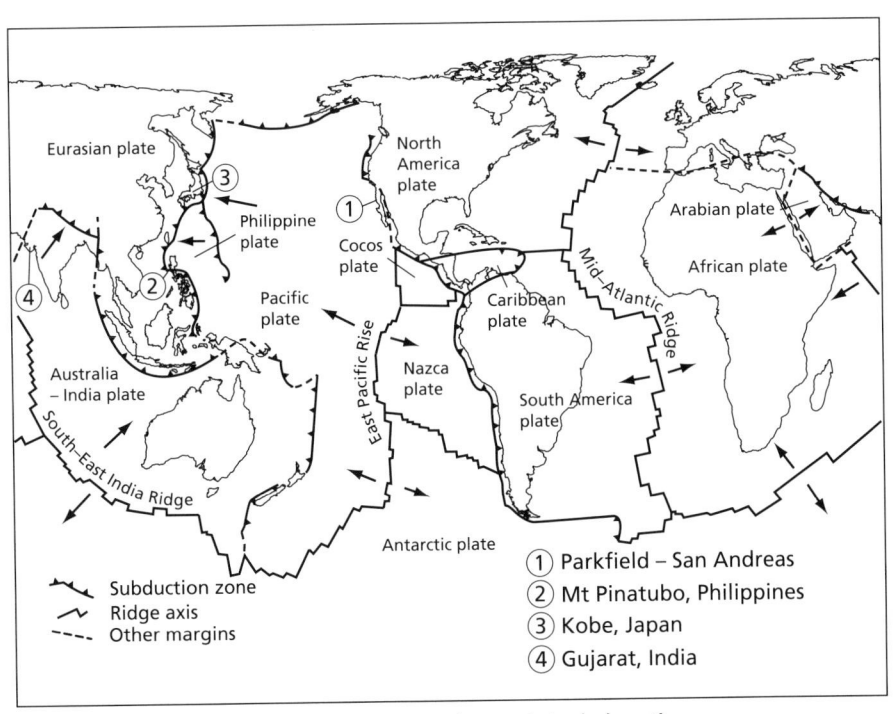

Resource 3: The major plates, their margins and study locations

Variations in the form of plate margins include;
- Destructive margins between oceanic plates – one plate will be subducted, causing earthquakes and volcanic activity. In this case volcanoes above the subduction zone will initially build as sea-mounts and may break the surface to form a line of volcanic islands – an island arc. The Pacific island chains of the Aleutians and the Northern Marianas are examples.
- Collision margins are created when plate-carrying continents meet. These may initially have been destructive margins but eventually the continental masses meet. No subduction occurs but rather the rocks of the land crumple, twist and fold to create high fold mountains. The collision between the plate carrying the Indian subcontinent and the Eurasian plate is the classic example, with the Himalayan fold mountains as a result.
- Conservative or Transform margins – in this case plates slide past each other. The best known example is in the Western USA along the San Andreas fault where a plate carrying most of California is sliding north relative to the rest of North America. As with most Conservative margins, earthquakes are common as stress builds over time and is suddenly released, but volcanic activity is less likely.

Confirmation of the theory of plate tectonics has come from numerous sources: the dating of ocean basin rocks; the depth of sediment on these rocks; the evidence of past magnetic shifts (paleomagnetism); and the empirical measurement of drift by laser. The theory is widely accepted and the details of the process are continuously being refined by research. There are over 700 active volcanoes globally and pinpointing of earthquake foci produces a map highlighting their location and the plate boundaries as they cross the oceans and continents (**Resource 3**). At these junctions plates collide, separate or slide past each other, producing distinctive landforms and events. Many have little direct impact on people – most earthquakes are in remote sea locations or are too small to be hazardous. Most volcanic eruptions are in isolated areas or build gradually giving warning. Nevertheless, the small proportion of earthquakes and volcanic eruptions that impact on people tend to do so in a dramatic fashion.

The hazards of volcanic activity

Each year about 50 of the world's volcanoes erupt, some for the first time in centuries (Pinatubo 1991), others as a regular event (Mt Etna, roughly every 10 years, 2001) while some are in almost continual activity (Kilauea in Hawaii).

Explosion

Some volcanoes erupt without significant violence, such as the volcanoes of Hawaii, but elsewhere the force of a volcanic eruption can be enormous. When Krakatoa erupted in Indonesia in 1883 it was heard 3000 miles away in Australia. The blast destroyed the island and 36,000 people drowned in the 40 m tsunami that swept the coasts of Indonesian islands. On one island a ship was washed 20 km inland. Tsunamis are huge waves generated by either volcanic eruptions or earthquakes, which can travel across oceans at great speed resulting in the devastation of coastal regions.

Materials released by volcanic eruptions

Numerous types of material are ejected by volcanoes including lava, pyroclastics and gases. In Hawaii and Iceland the local languages have dozens of terms for lava relating to its appearance and flow characteristics.

LAVA	PYROCLASTICS	GASES
Basic (runny) Acid (viscous) *Hawaiian examples* Aa (Ah Ah) *Blocky* Pahoehoe - *Ropy lava*	Volcanic Bombs Lapilli (small stones) Hot ash (fine dust) Cinders Pumice Nuee ardente (*glowing cloud*)	Steam Sulphur dioxide Carbon dioxide

Lava rarely threatens life as its flow is predictable, but it does destroy property by initiating fire or swamping buildings. In 1990 a lava flow in Hawaii buried the village of Kalapana. Over a three year period the flow gradually covered 181 houses, much farmland and the main coast road.

Nuee ardentes are spectacular, potentially lethal mixtures of superheated gases, hot ash and rock fragments. The death of over 30,000 inhabitants in the town of St Pierre, Martinique, in the Caribbean in 1902 is one of the best documented examples. Ash from the Mt St Helens eruption of 1980 entered the upper atmosphere and circled the globe helping create spectacular sunsets for months. Volcanic gases are often hot and toxic. When the town of Pompeii was uncovered centuries after it had been buried by an eruption of Mt Vesuvius in southern Italy, the buried inhabitants were found with their hands at their mouths or throats suggesting mass suffocation. In 1986 at Lake Monoun, Cameroon, over 1700 people died of CO_2 poisoning after gas issued from a nearby volcano.

Landslides
Volcanoes often bulge as magmatic pressure builds up beneath them. This deformation of steep slopes may cause landslides. Currently it is feared that an eruption in the Canary islands might cause a huge landslide to generate an enormous **tsunami** with devastating consequences, especially on the populous Eastern seaboard of the USA.

Lahars
These are volcanic mud flows. When hot ash mixes with river water or with heavy rain, which can be triggered by eruptions, it can flow as a thick, hot mixture at great speed flooding valleys, burying the environment and drowning people. In 1985 the eruption of Nevado del Ruiz in Colombia resulted in a lahar flowing at 140 kmph through the town of Armero some 50 km from the volcano. In one night over 20,000 of the town's 23,000 inhabitants perished, buried by hot mud.

Climate alteration
Major volcanic events, or a series of them, can impact the global climate. In 1815 the cold summer and consequent worldwide crop failures and famine in which millions died has been linked to the eruption of Mt Tambora in Indonesia. Scientists speculate that the disappearance of 90% of all species, including the dinosaurs, at the end of

the Tertiary era may be the consequence of a series of volcanic eruptions filling the upper atmosphere with dust and cutting solar insolation.

The benefits of volcanic activity

Benefits from volcanic activity include the fact that some, not all, lava flows can be weathered into rich, **fertile soils**. Soils formed from basic lavas and potassium or phosphorus-rich ash deposits are highly valued. Following an eruption in Alaska in 1912 the ash fall produced, in the following spring, the highest grasses and largest berry production ever known. It is no accident that over 20% of the population of Sicily live and depend on the mountain slopes of Etna despite its repeated eruptions, averaging one every ten years. Etna is at once an environmental, economic and social benefit.

Land

While ash and lava may bury useful land the same activity can create new land. The rise from the sea of Surtsey in 1963 near Iceland was a golden opportunity for scientists to study not only volcanic processes but also the development of a prisere and ecological succession. Another Icelandic eruption on the island of Heimaey, destroyed houses but the lava flows that had threatened the harbour actually enhanced the shelter provided for the local fishing fleet.

Mineral deposits

Volcanic deposits provide a wide variety of industrial materials and chemicals including sulphur, pumice, arsenic and boric acid. Mineral-rich gases within molten lava cool to form veins of ore such as the copper and tin deposits in old volcanic rocks of Cornwall. Igneous rocks themselves are prized as building stone, including granite. Gold and diamonds are volcanic in origin.

Energy

Reykjavik, Iceland's capital, gets most of its heating from geothermal water derived from volcanic springs. Over 50,000 homes receive water heated naturally to $87°C$ from this environmentally friendly system. In New Zealand, Italy and the USA naturally produced volcanic steam is harnessed to generate electricity. The largest such plant is The Geysers in California, generating 1000MW of electricity.

Tourism

There is an ancient link between recreation and volcanic activities. Hot water springs or spas across Europe have been a focus for travellers for many centuries. The Roman baths at Bath in England and Baden Baden in Germany are two examples. The health benefits of drinking or bathing in such hot and mineral rich water may be disputed but they remain popular. Volcanoes, especially active or recently active are strong magnets for adventurers and tourists alike. The 2001 eruption of Mt Etna in Sicily coincided with the holiday season and companies flew, coached and sailed thousands of visitors in to witness the event which was particularly spectacular at night. The mud pools and geysers of Yellowstone National Park, Wyoming, USA, are the key attraction for tens of thousands of visitors annually. The sheer beauty of volcanoes such as Mt Fuji in Japan is a priceless asset, and Crater Lake in Oregon is regarded as one of the world's most beautiful landscapes.

Chapter 2

Hazard prediction and management

Much research into floods, storms, droughts, volcanic eruptions and earthquakes is undertaken for the practical goal of minimising their impacts. Preventing natural hazards may not be possible, but accurate prediction of their occurance could reduce impacts on life and property. This is best appreciated where such events are likely, especially where large numbers of people and expensive infrastructure are found. Not surprisingly the leading experts on earthquake prediction are based in Japan and California.

Earthquake prediction

Recently a resident of Santa Cruz asked a news team why they were filming the main street and malls of the Californian city. The cameraman explained that they were shooting 'before' views – before the next earthquake, that was! In some locations predicting the 'Where?' is easy but the big question is 'When?' Californians await the 'Big One', an earthquake to match the 8.2 Richter scale event of 1906 when San Francisco either fell or later burnt to the ground. Theories of earthquake prediction mostly relate to either warning signs or seismic gap theory.

Warning signs

Tradition in China says that unusual animal behaviour is a prediction of an earthquake – ducks leave water, chickens fly into trees, dogs bark wildly and piglets eat their tails. No verifiable scientific evidence exists for this, but some suggest that gas or electrical emissions from the ground could be linked to such odd behaviour. In 1975 the Chinese authorities in Liaoning Province in Manchuria insisted three million people should sleep outdoors, despite the winter weather, on the basis of a major quake prediction. They were right. A Richter 7.3 earthquake struck. Tens of thousands would have died had they been inside the buildings that collapsed. As it was, only 300 were killed. This success was not based solely on animal behaviour. Rather the scientists had been monitoring earlier small tremors and tilting land measurements. Sadly, eighteen months later, elsewhere in China, a larger earthquake struck without warning leaving over 250,000 dead, possibly 750,000 as the actual figure was officially concealed. This was the worst earthquake death toll of the twentieth century and the hope of reliable earthquake prediction was dashed. In California the government has established an earthquake prediction centre in Parkfield, on the notorious San Andreas fault. Here the Earth's magnetic field, electrical resistivity, the level of gases such as radon and methane in the ground, the tilt of land and the water level of wells are all continuously monitored by sensitive hi-tech equipment. All these things are believed to change as a major quake becomes imminent. In Japan, after a 7.5 quake in 1964, records showed that the height of the coast adjacent to the epicentre had risen significantly for 60 years, only to drop rapidly in the days before the event. Such findings produced great confidence for accurate prediction in the 1970s, a hope that seems no closer today despite the major success at Liaoning.

Seismic gap theory
Russian and American scientists have produced very similar theories concerning the regularity of earthquakes. If these are correct, prediction may be possible. The underlying concept, at its simplest level, states that by monitoring earthquake zones and plate margins over time, a pattern of seismic events can be seen. If there are gaps along these zones – sections with little recent activity – then these are the likely stress building areas in which an earthquake should occur. Based on this concept, American scientists stated that a significant quake should hit Parkfield, California, between 1985 and 1992. It did not come and to date it has still not arrived.

Volcano prediction
Some volcanoes are highly predictable such as the Mauna Loa volcanoes, Hawaii, or the once-a-decade eruption on Mt Etna in Sicily. Others are much less readily anticipated. Of the 700 active (not dormant or extinct) volcanoes, only a fraction are monitored and only 70 are continuously surveyed in detail. Not surprisingly, volcanoes in the developed nations are studied more intensely and here prediction is achieved most readily.

Warning signs
These are similar to earthquake precursors: seismic events, tilting ground and gas release. Volcanoes by definition involve the release of magma and/or gas so beneath them material must be moving upwards causing earth tremors and bulging of the surface. Prediction can be accurate in terms of timing but inaccurate in terms of scale and impact. The 1980 Mt St Helens eruption was well monitored and predicted. A 5 km wide exclusion zone was set up and if the volcano had erupted vertically then it is possible no lives would have been lost. In the event, an earthquake beneath Mt St Helens caused a huge landslide on the mountain's northern flank, creating an outlet for the pressure, and an enormous blast of ash, debris and superheated gas erupted sideways devastating the landscape well beyond the 5 km zone in that direction. Sometimes it all goes wrong. In 1986 a Colombian volcano, Nevado del Ruiz, was monitored by scientists following signs of activity. After several weeks they declared that a major eruption was not imminent. The next day it did erupt and, as noted earlier, a lahar swept down an adjacent valley burying the town of Armero. It was little comfort to the scientists that they had accurately predicted the path of such lahars – only their timing was wrong. By contrast, scientists did evacuate many people from a threatening volcano at Mammoth Lake in California. No eruption occurred and the scientists faced the anger of residents over their inconvenience and economic losses.

Vulcanologists also monitor gas or lava emitted by volcanoes. Hi-technology laser monitoring is used to detect subtle changes in gases across the crater of a volcano. Changes in the chemistry or pressure of these gases may help to forecast the timing, scale or nature of future events. Since 1995, the Monserrat island volcano named Soufriere Hills has kept vulcanologists busy assessing the nature and the future of its on-going activity. Here remote sensing from satellites has been employed to detect thermal alarms and to plot lava flows or fumerole emissions using UV enhanced camera shots.

Case study: Predicting the 1991 eruption of Pinatubo, Luzon, Philippines

First signs

In July 1990 a 7.7 magnitude earthquake struck the Philippine Islands in South East Asia killing about 1600 people. A natural disaster itself, it was also the first indication of another major event less than a year later. Unknown to anyone, in the months following the earthquake deep beneath the surface, the subducted Philippine plate was melting forming magma. This molten rock rose from the upper mantle or asthenosphere into a vast magma chamber beneath a 1759 m peak on the island of Luzon. The hot intrusion reactivated material in the chamber creating a mass of gas-charged magma. This mass continued to rise up through the crust of the Eurasian plate towards the summit of a long dormant volcano known as Pinatubo (**Resource 4**). In April 1991 Aeta villagers noticed steam issuing from side vents in their neighbouring mountain that had not shown any life in 600 years. This was reported to local officials, and scientists from the Philippine Institute of Volcanology and Seismology (PHIVOLCS) in Manila were brought in. Their leader, Dr Ray Punongbayan, ordered an aerial survey of the area and initially the report suggested the mountain was merely 'letting off steam'. A portable seismometer was installed and over 400 earthquakes were recorded in two days. Punongbayan contacted the United States Geological Survey who were not only interested in the volcanic activity but also in protecting two large US bases then in the region, the naval base at Subic Bay and the Clark Air Force base. By the last week of May, American scientists, including some with experience from Mt St Helens in 1980, were working with the Philippine team. They set up their headquarters in Clark Air Force base (PVO – Pinatubo Volcano Observatory) and established a network of seven seismometers around the mountain.

Watching and waiting

The next seven weeks was an intense period of study and debate. Initial recordings quickly dismissed the less dangerous possibilities of a simple steam release by the volcano or purely tectonic activity beneath it. Magma was rising. The questions were how much and how far would it rise?

One of the best guides for vulcanologists is the history of the volcano. Pinatubo had last erupted in the fourteenth century and no written record existed. A field study of the deposits and flows of pyroclastic material on Pinatubo's flanks showed it had erupted four or five times in the last two millennia. Such infrequent events are usually violent eruptions and the flow deposits confirmed this explosive nature.

Resource 4: The location of Mt Pinatubo, Philippines

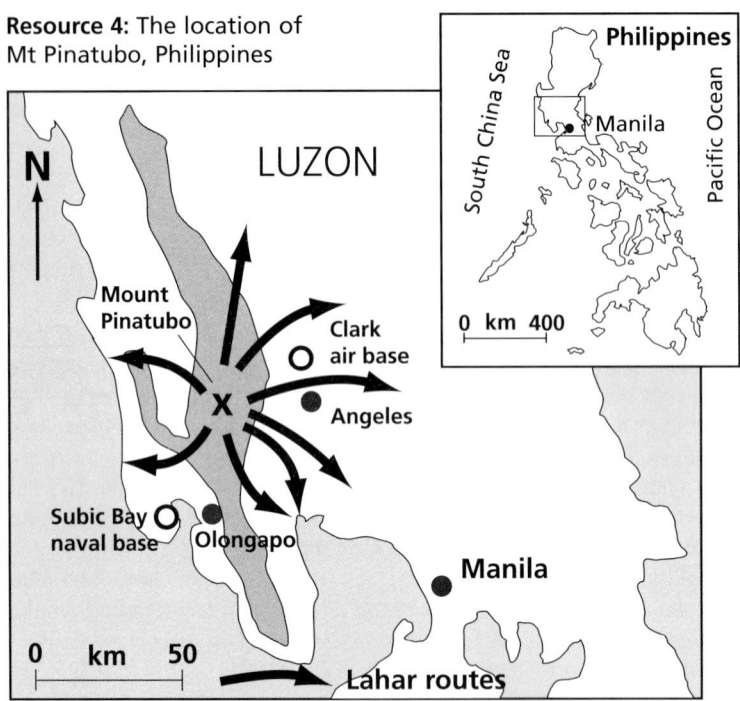

In addition to the continuous monitoring of the seismic net, daily flights by plane and helicopter were made. Gas samples were taken using a Cospec (Correlation spectrometer). In particular the sulphur dioxide (SO_2) levels emitted from the vents were studied. The first record of 500 tonnes per day was significant but this rate rose over the subsequent five weeks to 5000 tonnes. Meanwhile, earth tremors at depths of 8 km continued at a rate of between 40 and 150 a day. This was not an unusually high rate but their spread over 5 or 6 km indicated rising magma over a broad area.

The end game

On 2 June the activity increased and the largest earth tremors yet recorded occurred the next day. At PVO the teams decided to issue their first alert at level 2 (see below). Twenty thousand people living within 10 km of the mountain were evacuated and the 15,000 personnel at Clark base were told to prepare to leave at short notice.

Table of levels of eruption alert

LEVEL 1 -	Activity detected	Eruption NOT imminent
LEVEL 2 -	Activity more intense	Eruption probable
LEVEL 3 -	Eruption now likely	Possibly within 2 weeks
LEVEL 4 -	Eruption possible	Within 24 hours
LEVEL 5 -	Eruption in progress	

On 5 June the earthquakes were more concentrated beneath the central zone of the mountain and the sulphur dioxide levels were falling. This second factor was a significant one suggesting that the magma was now retaining its gas content, making it potentially explosive in nature. But the events were waxing and waning in intensity and the team were very concerned about getting the timing right. The next day, 6 June, a slug of fresh lava was seen and a level 3 alert declared but the bulge of apparently fresh lava was later classified as an older remnant and fears of a false alarm rose. Next morning building seismic activity prompted the scientists to declare a level 4 alert and 120,000 people living within 18 km of Pinatubo were moved to temporary evacuation centres. The air force base personnel remained but on alert to move.

At this point reports arrived of a volcanic eruption in Japan. Mount Unsen, which was also being carefully monitored, had erupted as expected but with much greater violence than predicted. Six thousand locals had been evacuated and saved but 34 individuals were killed by an avalanche of hot gas, ash and rocks. It was a timely reminder for the team at PVO.

On the 8/9 June a dome of thick, relatively cool lava was spotted from the air. Such material matched with the old deposits from previous eruptions and confirmed the likely violent nature of any eruption. The US air force base was evacuated the next day leaving only a security staff of 1500. A 60 km wide danger zone (not evacuation area) was declared, including the 300,000 population of the city of Angeles. Only 48 hours later the first major eruption occurred. It covered the dense forest around Pinatubo in a thick layer of grey ash. For 24 hours large eruptions continued with ash scattered up to 80 km away. At 2 am on the 15 June the first of a series of five massive overnight explosions occurred. A vast cloud of debris stretched over 16 km across and up over 30,000m into the atmosphere. Pyroclastic material was scattered in all directions and well beyond the exclusion zone. These massive eruptions released more material – a total of over 20 million tonnes – into the atmosphere than any other 20th century eruption. To compound the problem, the eruption coincided with a typhoon with hurricane force winds and intense rainfall. The mixture of rain and ashfall buried thousands of hectares of farmland, killed over one million livestock and caused thousands of buildings to collapse under the sheer weight of debris. Schools, hospitals, children's homes and houses were destroyed in the city of Olongapo, 56 km south west of Pinatubo, where many evacuated groups had moved (**Resource 4**). The final death toll was around 900 and costs were estimated at £10 billion. Mount Pinatubo continued to erupt sporadically during 1992 and 1993 before finally settling perhaps for another 500 or 600 years of dormancy.

Conclusion

The work of prediction by the scientists at PVO was over and their accuracy undoubtedly saved thousands of lives. In the months after the eruption the wet monsoon generated numerous lahars of fast-flowing water and hot ash. Some of the local vulcanologists turned their prediction skills to address the problems of where and when these events would occur.

Managing and responding to earthquakes

Earthquakes effects are commonly divided into Primary (Short Term) and Secondary (Long Term).

Primary effects of earthquakes
- Death and injury.
- Human fear and anxiety.
- Buildings collapse, wholly or in part, burying or trapping people.
- Other structures collapsing: bridges, flyovers or elevated routeways.
- Phone, road, rail and other communication links disrupted.
- Fracture of underground services such as water, gas and sewage pipelines.
- Fires may be started or made worse by gas leaks.
- Landslides, shifting and cracking ground.
- Liquefaction - a process whereby soft or unconsolidated sediments amplify the effect of shaking ground and act as a liquid often causing building foundations to sink or subside.
- Out at sea tsunamis are formed; these are fast moving high sea waves that radiate away from undersea quakes. They often have a devastating impact on coastlines.

Secondary effects of earthquakes
- Homelessness and lack of adequate shelter.
- Long term communication problems which may delay supplies of food and medical assistance.
- Vital services are disrupted and may cause a lack of clean water and/or sewage disposal.

Note that all these effects may cause deaths, as well as illness and increase misery for the survivors of the initial impacts.

- Bereavement, family disintegration and parentless children.
- Stress and long term emotional problems.
- The huge cost or debt in rebuilding infrastructure.
- Loss of jobs, closure of businesses and factories.
- Refugee camps may exist for months or years and the people may leave the region altogether.

Any or all of these impacts may follow an earthquake but their intensity and severity may have less to do with the magnitude or nature of the earthquake than with the country or region involved. In general terms more advanced nations in scientific and economic terms are better placed to prepare for and respond to an earthquake episode. Knowledge and perception are critical in making arrangements for earthquake response. Despite years of research we are no closer to making accurate predictions about earthquakes and even the worldwide distribution of these phenomena is incomplete.

Earthquake study 1: Kobe, Japan 1995

The event
On Tuesday 17 January 1995 an earthquake measuring 7.2 on the Richter scale occurred along a 50 km section of a complex plate margin off the coast of south-central Japan. An 8 km deep ocean trench marks a subduction zone between the Pacific, Philippine and Eurasian plates. The focus of the earthquake was 30 km below the surface and above it the epicentre lay adjacent to Japan's second largest concentration of population. Over ten million people in the cities of Kobe and Osaka were impacted by the 20 seconds of ground shaking (**Resource 5**).

The primary effects
The death toll came to 6300 with around 35,000 people injured. In the first day 30,000 were made homeless as 180,000 buildings were destroyed or badly damaged. Several highways collapsed including a 500 m stretch of the Hanshin expressway where the Shinkansen bullet train track was twisted and services halted. Over 150 fires initiated by gas leaks or tumbling paraffin stoves swept through traditional wooden homes, many of which had collapsed in the first minute of the event. Kobe's port, the largest in Japan was devastated. Nearly 200 ship berths were destroyed including many on a recently reclaimed island. Here liquefaction in the soft sediments heightened the impact.

Resource 5: The location of Kobe, Japan

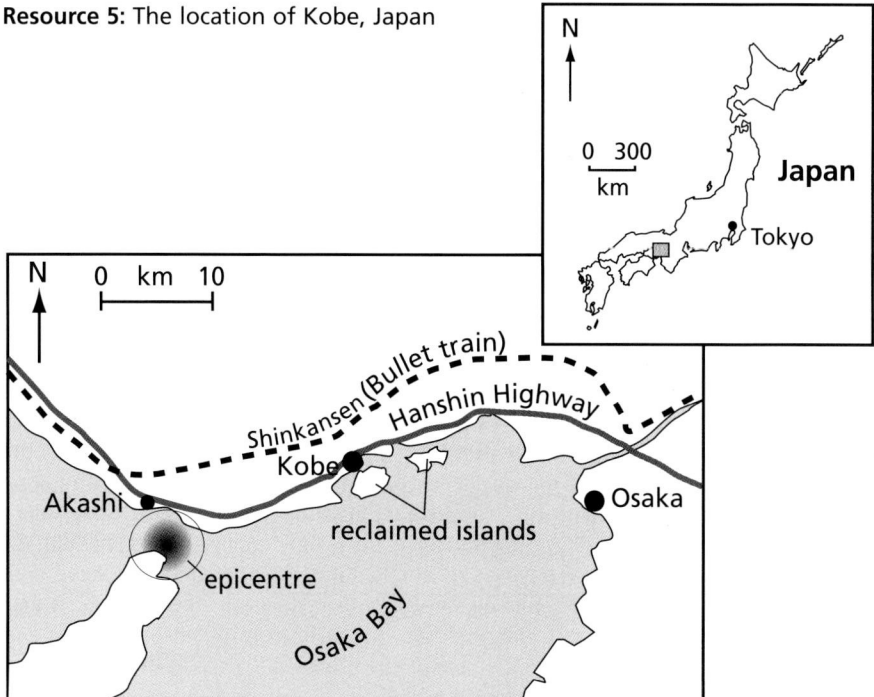

The secondary effects
The number of homeless rose to 310,000 in the aftermath of the fires. This figure included 30% of Kobe's resident population moved to temporary shelter in camps. Rebuilding costs were estimated at £70 billion, while the total cost of damages, including lost business, was set at over £120 million. As only 3% of the population had earthquake insurance, the financial losses had enormous impact on individuals. Gas supplies remained off for up to three months due to severe underground damage. Local medical services reported continuing emotional stress amongst survivors and the bereaved in the months following the earthquake. About 96,000 people moved away from Kobe, reducing its population to 1.4million.

Knowledge, perception and stage of development
Japan has a worldwide reputation as a nation that is acutely aware of the threat of earthquake damage and one that has taken steps to prepare for such an event. The Kobe earthquake, named by some the 'Great Hanshin', provided a test of these measures and highlighted their limitations. As with most quakes it was buildings that created the high death toll. Many buildings fell or suffered from 'pancaking'. This is when poorly supported upper stories fall onto those below. In the centre of Kobe nearly all of the modern steel-framed office blocks survived though some innovative designs did not perform so well. Research shows that structures, including bridges and elevated highways, built in the 1960s and 1970s were subject to high levels of damage compared to those constructed after 1981 when new building standards were introduced and, more significantly, enforced. These new regulations implemented the use of stronger materials and more flexible structures that could sway and bend with seismic waves. In reviewing the immediate response to the earthquake it was found that both the emergency response and the search and rescue operations were unsatisfactory though many lives had been saved. The preparations covering school buildings and street signs marking escape routes had performed well. Less successful was the investment of £70 million each year on earthquake prediction. No warning signs were recognised although it was noted that the area had had a lack of events over the previous 50 years, hinting at the build-up of stress. The scientists were convinced that the next likely major quake would be in another region. One researcher studying groundwater content and level did record, on 7 January, radon-222 gas levels twelve times above the normal in wells near Kobe. Unfortunately this information was only highlighted months after the incident.
Perhaps where Japan's knowledge and level of development is most noteworthy is in the recovery after the event. On the day after the earthquake, apparently many shop keepers opened for business on the pavements and within 12 months 70% of Kobe's shop and businesses were fully functioning again (some others had relocated). By October 1996 the Hanshin highway was fully rebuilt and the port was 70% operational within a year and completely restored by 1997. In conclusion, the earthquake has encouraged new resolve and urgency on strengthening older structures and maintaining tight building controls but Japan still grows and, despite the liquefaction threat, reclaimed land continues to be developed.

Earthquake study 2: Gujarat, India 2001

The event
On 26 January 2001 at 8.46 am an earthquake registering 7.6 on the Richter scale occurred along a fault line in the Rann of Kutch, which lies in northwestern India near the Pakistan border (**Resource 6**). The focus was about 17 km below the ground surface and the epicentre was near Bhuj, a town 200 km northwest of the region's largest city, Ahmadabad. The quake was felt as far away as the cities of Delhi and Mumbai (Bombay). Significant aftershocks continued for a month. The region is not adjacent to a plate boundary but it has a long history of earthquakes at a series of crustal faults. The last major event was a Richter 7.0 quake in 1956.

The primary effects
Death tolls in earthquakes are notoriously inaccurate and here initial reports were up to 100,000. A month after the event the official figure was 19,727 but this was revised down to around 13,000 as bogus and double recordings were eliminated. The injured were estimated at 166,000 and over 600,000 homeless. A third of a million houses were destroyed and nearly a million more badly damaged. Transport was severely disrupted

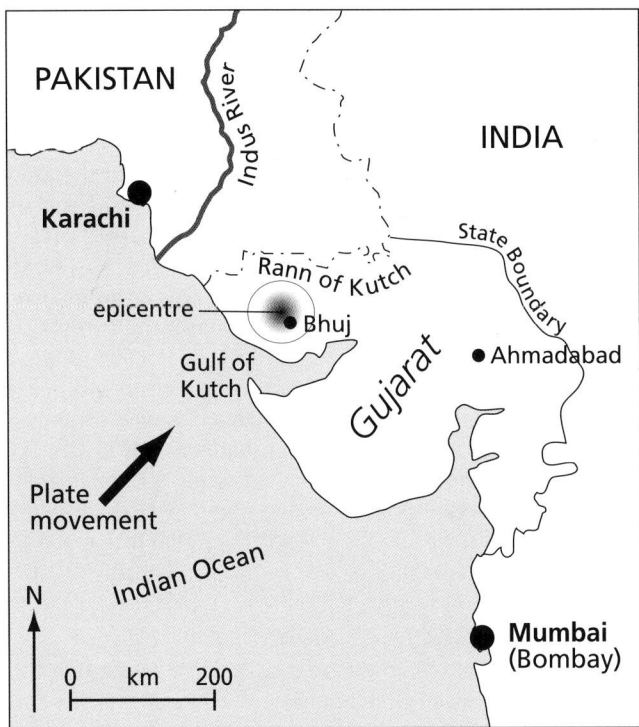

Resource 6: The location of Gujarat, India

with the key Surajbari bridge linking Kutch and the rest of Gujarat destroyed. Fibre-optic cabling and the telecom building in Buhj were destroyed, further isolating regions and hindering the rescue operation. The direct economic toll was stated by government sources as $1.3 billion though others suggest a higher real impact in the order of $5 billion. The quake struck on India's national holiday, Republic Day, celebrating the establishment of their republic in 1950. As a result, schools and government offices were closed and this may have contributed to the high casualty figures. Most victims were women and children and the vast majority died as the result of building collapse. Villages and towns were wiped out as houses were turned to rubble. Bhuj and Ahmadabad recorded thousands of deaths as buildings up to ten stories high pancaked. The central Bhuj civil hospital was razed to the ground. One town illustrates the problems which faced numerous similar settlements. Bachao, population 42,000, was a prosperous town with around 1000 buildings. It lay about 100km from the epicentre. Nearly all its buildings collapsed including six salt factories and their worker hostel accommodation alongside. A small 20 bed hospital full of patients and staff fell within seconds of the tremor. Bridges, petrol stations and roadways were destroyed as the ground heaved and fissures opened. No outside help came for days after the event and locals used their bare hands in a desperate search and rescue effort. When rescue teams finally reached Bachao, numerous funerals pyres were burning across the town.

The secondary effects

The rescue operation was hampered by many factors including the continuing aftershocks and the disrupted transport and communication networks. The Indian army, including six engineer regiments, navy and air force were mobilised and many international teams came to help. These included a Swiss team with trained sniffer dogs. Medical field hospitals were set up using mobile generators and in the week after the event over 10,000 surgical operations were performed. Newspaper articles highlighted local government failings as many rescue workers wasted time waiting to be allocated to problem sites. A lack of coordination was widespread. Being January, cold winter nights were a real problem for the vast number of homeless. Blankets and tents were urgently required; frequently the number supplied was a fraction of what was needed. The Red Cross found many villages had received no such emergency relief supplies a week after the quake even though supplies were piled up for days on airport runways waiting to be transported. Medical experts feared outbreaks of disease as bodies decomposed and water supplies became contaminated. Reports came in from doctors of rising illness including diarrhoea especially among young children. A feared epidemic of cholera did not occur. Thousands of refugees left the Kutch region due to the fear of aftershocks and chronic shortages of food and water.

Knowledge, perception and stage of development

The Indian Prime Minister, Vajpayee, admitted, in the aftermath of the quake that the country was not ready to face such a disaster. The local Gujarat government took two days to set up a control room and even longer to move in appropriate heavy earth-moving equipment. Officials faced a barrage of criticism, some suggesting they were ignoring the region by not visiting problem areas, while others pointed out that the PM's visit entailed the closure of roads used by rescue workers and the suspension of emergency flights. Within a week of the disaster many local builders were under investigation by the police following allegations of criminal negligence with respect to building codes of practice. In Buhj it was admitted that buildings up to eight stories high had been built without any checks or inspections in the two years before the event. Even the state buildings did not meet earthquake standards and while the local nuclear power plants and dams survived, the Buhj state hospital and much housing was destroyed. Officials had largely ignored a 1998 study of the region that noted that 'Disasters don't kill people, buildings do', and '... the number of unsafe buildings is increasing daily'. A press report from Ahmadabad stated:

> Bribing officials to overlook poor workmanship or code violations is not uncommon, and officials are often unqualified to carry out inspections. A survey of engineers after the earthquake showed that in many cases, the pillars were not secured to foundations with steel reinforcements ... When the earthquake struck, many such buildings swayed and collapsed under their weight or because walls of cement buckled.

The rebuilding of housing and compensation for victims is an ongoing problem with continuous delays and complications. In one area the relocation of villages has been hindered both by a debate over the precise location of the epicentre and by the conflict between the upper and lower ranks of the caste system.

Indian and foreign scientists have studied this region for decades and have been aware of its active seismic nature including an earthquake registering Richter 8.0 back in 1816. This knowledge on paper has not been translated into practice in either preparedness for, or response to, such an event. Short-term progress had prevailed over long-term safety. For this, numerous local builders now face fines and imprisonment but many argue that it is the officials together with local and central government authorities that should be made to answer. As journalist Deepal Jayasekera wrote after the quake, '... it is the profit requirement of business – not the social need of the masses – that determines government priorities before, during and after such calamities'.

Chapter 3
Extreme weather in mid-latitudes causing hazards for human activity
Prolonged drought of the summer of 1995

Introduction
Weather hazards in mid-latitudes are generally a rare occurrence. Without the heat energy that is available in the Tropics the low-pressure systems, for example, do not have the same intensity as, or power of, their big brothers – the hurricanes. The position of the UK in temperate latitudes between 50 and 60 degrees north means that for most of the time throughout the year, its weather is dominated by frontal low pressure systems or depressions which arrive in from the Atlantic. These are tame affairs and seldom create hazards for the people living in this zone. Severe storms experienced by the UK, like those of October 1987, January 1990 and December 1998, have been the exception rather than the rule. I hope to provide further information on a case study of the 'Burns' Day' storm (25 January 1990) on the NINE website.

High-pressure systems, both in summer and winter, can present a set of different but challenging hazards and in this chapter I want to focus on a prolonged drought. The main cause of prolonged spells of dry, warm weather over the UK is high-pressure systems in the summer months such as that established during the summer of 1995.

Causes of prolonged dry periods
An anticyclone is a high-pressure system, which develops when upper air streams undulate, pushing air downwards. Some of this air finds its way to the surface creating a stable dome of air on the surface. The majority of anticyclones are simple ridges of high pressure sandwiched between frontal depressions and last only a day or so. Occasionally, these high-pressure systems remain stationary over the UK for longer periods and act as a block to the path of low-pressure systems. When this happens the weather remains stable for several days if not weeks. These are known as **blocking highs** and caused the prolonged weather pattern found over the UK in the summer of 1995.

The weather associated with anticyclones may vary according to the time of year. A prolonged high-pressure system in summer produces very different climatic conditions to that in winter. In summer this type of system creates very warm, dry conditions with little wind and cloudless skies. The anticyclone produces dry weather because the air is sinking and as it descends, the air warms up. This happens because the air is compressed into a smaller volume of space as it descends closer to the surface. The air warms up at a rate of approximately 1°C for every 100 metres. This is normally referred to as the dry adiabatic lapse rate – the rate at which dry air heats up or cools down when moving vertically through the atmosphere. This warming process also enables the air to hold and retain more moisture and, with little or no upward movement of air, no condensation takes place and therefore cloud formation is inhibited. The light winds that are produced tend to blow in a clockwise direction out from the centre of the high.

In winter, anticyclonic weather is somewhat different. The main difference is that the air is colder and this leads to condensation of moisture near the surface within the system. The combination of low rates of insolation and shorter daylight produces colder weather. However, as we are concentrating on the issue of extreme weather associated with drought, no further discussion of anticyclonic weather in winter is called for.

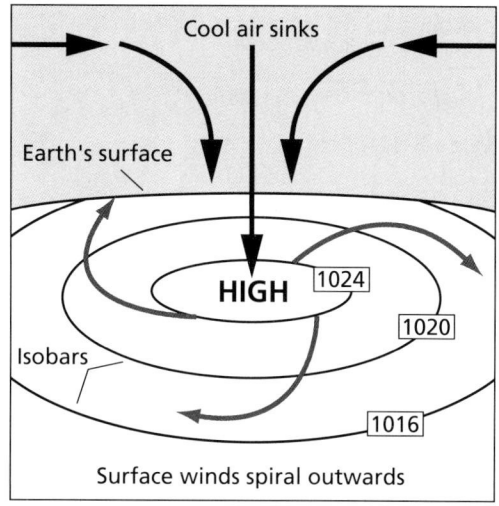

Resource 7: A high-pressure system

Case study – summer 1995

Anticyclonic systems dominated the weather in the months of June to August 1995. It was the driest since records began in 1727. The month of August was characterised by hot temperatures (between 3.5 and 3.8^{0}C higher than the seasonal norm), high rates of sunshine (55% above normal) and low rainfall totals. The exceptional warmth began in the last week of June and lasted until the end of August. The total precipitation for England and Wales was only 47 mm compared to the 1961–90 average of 139 mm.

During this summer a number of high-pressure systems established themselves over the UK and remained stationary for those months. The position of the anticyclone was crucial during that summer. The blocking high stabilised over the centre of the UK or to the north-east over the North Sea. As the winds rotated in a clockwise direction around the centre, warm air over Europe was drawn in to the UK from southern regions maintaining the dry, hot weather being experienced over the whole country.

Impacts and human responses to the drought of 1995

There were both positive and negative impacts from the warm year. Firstly an examination of the impacts on the natural environment is required before appraising the impacts on the human sectors.

Impacts on the natural environment

An important source of pollution associated with hot weather is the appearance of blue-green algae in watercourses. While they are breaking down, toxic chemicals are released which can adversely affect humans and animals. Algal blooms significantly

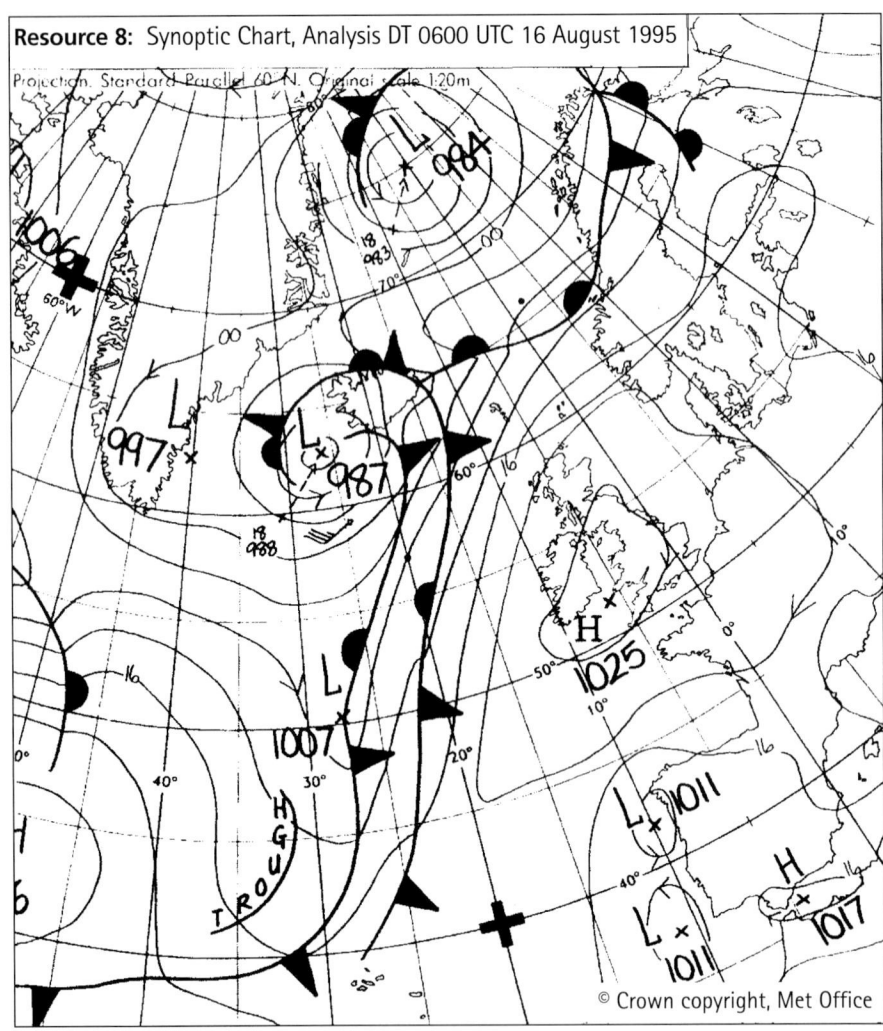

Resource 8: Synoptic Chart, Analysis DT 0600 UTC 16 August 1995

© Crown copyright, Met Office

affected the River Great Ouse in England. Overall however, although some recreational and tourist facilities had to be closed, the level of pollution on most rivers was not as great as expected. This was due to better monitoring and management practices, which reduced the amount of nutrients reaching the watercourses.

Air pollutants that were influenced by hot weather during this summer included ozone and particulates (PM_{10}). Ozone in the lower atmosphere is produced by a complex set of reactions (of hydrocarbons and nitrogen oxides) and sunlight. During this period of hot weather, ozone levels were lower than expected from the temperature relationship. This it is thought was due to the relatively clean air masses arriving in the UK from areas over the sea rather than from parts of central Europe, as was the case in the 1976 drought. In terms of particulates, there is a distinct seasonal pattern in concentrations of PM_{10} during the hot weather. These concentrations occur principally due to the lack of a dispersion mechanism, ie wind, and with little, if any wind associated with high-pressure systems, the concentrations tend to be higher. Indeed the concentrations exceeded safety levels (50 mg/m^3 over a 24 hour period) for a number of places. These can have serious health effects because, given a high enough concentration, they can cause respiratory problems and even death.

The amount of tree growth, especially among shallow-rooted trees in open country environments, such as beech, ash and sycamore, was curtailed during the dry summer of 1995. Deep-rooted species such as oak were generally unaffected by the drought. The number of outdoor fires in forested areas was significantly up on 1994 – the last year with a long dry period. The dryness of the timber and the increase in people involved in outdoor recreational activities were to blame.

Impacts on the human systems

Agriculture

The most obvious impact was seen in the agricultural and water supply sectors. There were substantial losses in revenue and increased costs respectively in these two sectors. The impacts on arable crops are estimated at a financial gain of £30 million, while there was underproduction and low quality resulting in losses to other crops such as potatoes and vegetables. Overall, crop yields did better than the dry year of 1976 because the soil moisture content remained higher in 1995 and there was greater use of irrigation in many areas. The impact on livestock was in the form of increased costs for purchased feeds and a decline in the fertility of pigs and poultry. Farmers responded by adjusting their management techniques to take into account the low availability of water. This meant greater irrigation for arable crops, sprinkling and dousing equipment for livestock and improved water quality for aquaculture (fish farms).

Water supply

The exceptionally warm weather in 1995 had economic and regional implications. The dry weather led to widespread water shortages and costly drought relief operations in some areas. The rainfall total for England and Wales was only 9 mm for August and 38 mm for July. By July the Water Services Association had identified a serious water supply problem for many regions in England and Wales and by early August there was a hosepipe ban in operation in many places. Oddly enough the regions most affected by the drought were located in the north, which normally had the most rainfall. By late August, water stocks in major reservoirs in the Pennines and the Lake District had dropped to one fifth of the capacity. Given that most of the water companies admit that one quarter of water supplies are lost due to leakage from the transmission system (pipes), there was a serious issue developing during this summer. Overall, the financial impact of this hot summer for the water service companies has been estimated at approximately £96.1 million extra. In terms of the cost of supplying water, the hardest hit regions were the north-west and Yorkshire, which it was calculated incurred exceptional costs of £24.4 million and £47.0 million respectively.

Retailing
Weather has long been recognised as having a major influence on certain types of goods and services. Pioneering work in America first highlighted the weather-retail relationships such as making it difficult, or even preventing, people going into shops (snow drifts), or influencing the type of products purchased. Most importantly, weather has a financial impact on retailing. The influence of sunshine and temperature has a significant negative impact in summer because people prefer to participate in outdoor activities rather than go on shopping trips.

This is truer now than it was during the drought of 1975 (Agnew, Palutikof, 1997) because people have greater mobility today and can go to the seaside or countryside rather than do the shopping. Additionally, with almost universal refrigeration available in households, shopping trips can be deferred or postponed. The strongest impact was the sale of footwear and clothing. Sunny weather discouraged shopping for such items and losses in this sector were estimated to be in the region of £380 million. On the other hand, it is not surprising that demand for drinks, particularly light alcoholic drinks and soft drinks, grew continuously during the summer months of July and August.

Tourism
Tourism plays a significant role in the economy, earning approximately £100 million per day from all visitors. Internationally, the UK is the sixth most popular destination for tourists who are attracted by its heritage and history. The summer of 1995 produced record levels of tourists (23.7 million) visiting the UK and that, plus the large number of people remaining at home to take their holidays, had a large positive financial impact on the UK economy. The number of trips and occupancy rates for hotels and boarding houses both increased significantly during the summer months. Traffic jams in certain tourist areas such as the South West were an irritating drawback experienced by many holidaymakers during the summer months.

Human responses
The hot summer of 1995 was marked by numerous and persistent algal blooms along rivers. In response, most authorities banned any recreational activity on the affected parts of the river, advised the public to avoid contact with the water and told farmers to ensure their livestock did not drink from the river. Such a hazard happened to a 15 mile stretch of the River Great Ouse in East Anglia where water levels were unusually low. This did cause a lot of frustration among members of angling clubs in the area due to the loss of the opportunity to fish in the warm summer.

Responses to the water shortage targeted the customers initially, and secondly the supply system. The responses to excessive customer demand were to shut down supplies and open up standpipes in the street, impose hosepipe bans to stop unnecessary watering of gardens, reduce irrigation on farms and import lorry-loads of fresh water supplies to towns from other areas. Drought orders were imposed in a number of regions to ration the supply of water in these areas. The North West relies a lot on surface water resources for its water and with the drought came low volumes in the reservoirs. In 1995 the rainfall was only 50% of the 1961–90 average figures yet a peak in demand was reached in early August. This did reduce in response to an appeal to conserve water supplies. It revealed a need for more investment in water resources in this region to meet peak demands in times of drought.

The response to the supply side of the equation tended to happen after the event and took the form of public enquiries into the various companies and systems supplying the water. There has been a progressive improvement in water supply infrastructure since 1995.

In agricultural regions such as East Anglia, the demand for spray irrigation increased with the dry conditions experienced during the summer of 1995. Root crops in particular required extra water and this in turn put stress on the limited water resources in that area. The long drought had a negative financial impact in many agriculture areas because water resources were restricted and, as a consequence, commercial yields were down. The response has been an increase in long-term investment in water supply infrastructure to avoid such difficulties in the future.

Retailers are now acutely aware of the seasonal impact of weather on different sectors of the trade. Clothing and footwear are clearly affected during hot weather while the drinks industry benefits directly from the hot weather. Many clothing retailers now rely on the weather forecast much more than before to enable them to react quickly by ordering in appropriate clothes to match the weather predicted. Such has been the demand that the Met Office has established a commercial unit offering specialised weather advice to retailers in weather-sensitive sectors.

The response of most people to the hot weather was to remain at home for their holidays. This had a significant economic benefit to many coastal resorts around the UK. It enabled some of them, eg Blackpool, to invest in new attractions and upgrade their facilities in terms of types of accommodation available.

In conclusion, the summer of 1995 produced a number of significant impacts in diverse sectors, both natural and human. Environmental systems seemed able to recover quickly from the hazards and generally suffered minimal damage. The hazards for the human systems were much more variable, but no long-term damage was sustained beyond one year. Most of the outcomes were judged on the economic impact this had on the regions or the country as a whole. Beneficial outcomes were obvious in sectors such as tourism and in certain retailing sectors, while in agriculture and public water supply extra costs were incurred and impacts were deemed to be more negative. Transport systems seemed to come out reasonably well with the costs and benefits balancing each other out. Human responses also varied, from the minimalist approach evident in the agriculture sector to a more radical review of systems processes, initiated by the water supply sector.

Overall, this summer produced a number of hazards, principally in the form of economic problems. With few exceptions, most of these hazards caused human systems in the UK to bend rather than break. However, if the threat of global warming is real, such systems will have to be even more flexible to cope with many summers similar to this one in the future.

Chapter 4

Monsoon airflows

The monsoon derives its name from the word 'mausim' meaning 'season' and is the reversal of pressure and winds which gives rise to a marked seasonality of rainfall over south and south-east Asia. It is amongst the best known climatological phenomena in the world with countries such as India, Bangladesh, Philippines and northern Australia experiencing it.

This climate can boast of producing the wettest place on Earth – Charapunji in Assam with 7,000 mm per year! Each monsoon region has its own particular character, and none more so than the monsoon climate of India, which will form our study region. In simple terms there are two seasons – the hot dry season (north-east monsoon) experienced from November to May, and the hot wet season occurring from June to October (south-west monsoon) as can be seen in the climate graph (**Resource 9**) for Mumbai, formerly Bombay. What makes this climate so noteworthy are the stark differences between the two seasons in terms of the wind and rainfall patterns. Monsoon seasons do not consist of a period of rain falling continuously but rather a few days of torrential rain followed by a respite of sunshine to be followed in turn by more torrential rain. Historical evidence suggests that this climate developed with the formation of the Himalayas, which began 15 million years ago, but had reached a height about six million years ago whereby they began to interfere with existing atmospheric circulation. It is estimated that 70% of the rainfall has its source in the Indian Ocean. Other sources of rainfall include depressions originating in the Bay of Bengal which track north-westerly across central regions and, of course, cyclones which are

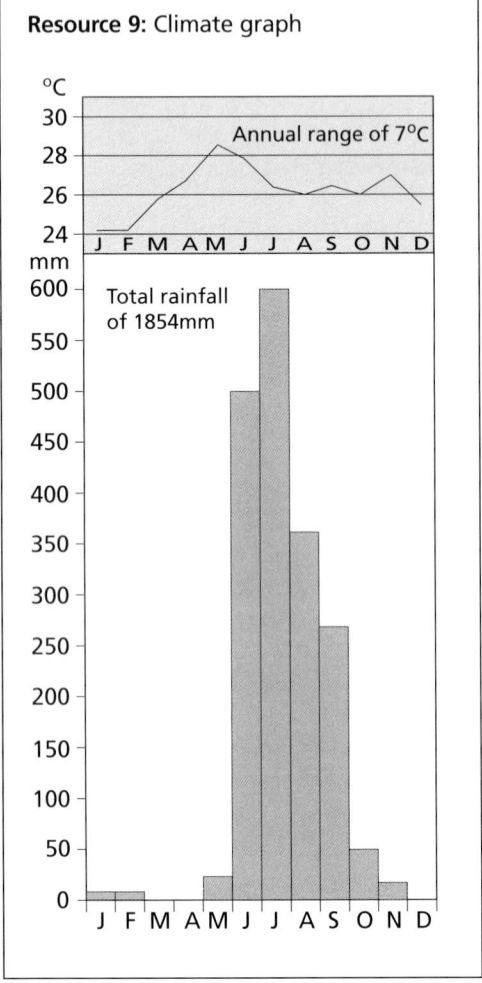

26

intense tropical storms that form in the Indian Ocean or Bay of Bengal. Cyclones are commonly experienced between late September and early December and can bring great destruction to coastal regions of India and Bangladesh, eg September 1988, October 1998.

Causes of the monsoon climate

The basic causes of the Indian monsoon have been known for many years but detailed knowledge of specific processes have only recently come to light with better data and more sophiscated technology. New data generated by radiosonde experiments on the vertical structure of the atmosphere helped produced a new understanding of the processes that caused the monsoon reversal of winds. A number of factors need to be considered when explaining the different airflows that produce the monsoon.

- The presence of a large landmass, which heats up rapidly with the approach of the overhead sun, and the existence of a large ocean nearby produces extreme pressure differences at different seasons. These differences constitute a steep pressure gradient which in turn produces winds that blow from different sources at different times of the year.
- The heating or cooling of the continental area is directly related to the movement of the **Inter Tropical Convergence Zone** as it follows the pattern of the overhead sun throughout the year.
- As the winds blow from high pressure to low pressure they are deflected by the **Coriolis force** which affects which regions receive the monsoon rains and when.
- The presence of different **jet streams** – the westerly jet over the Himalayas and the easterly jet off the south coast of India – are capable of transferring a large amount of air in and out of the region and help sustain continuous surface airflows at different times of the year.
- The presence of the **Tibetan Plateau and the Himalayas** prevents the normal smooth shift of the pressure belts with the seasons. Instead they help generate an abrupt shift of the westerly jet to the north of the Himalayas causing the sudden onset of the monsoon.

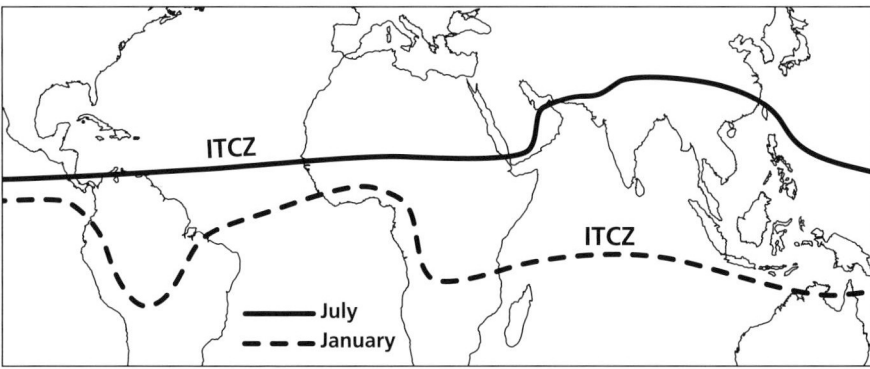

Resource 10: Changing position of the ITZC

- **Topography** plays an important role in determining amounts of rain received. Some high-altitude areas such as the Western Ghats force the moist air to rise and cause relief rainfall. In other regions whose slopes face the SW, eg the Assam hills, there is increased precipitation because they lie in the track of the monsoon depressions and close to the Bay of Bengal.

The South West Monsoon (summer monsoon)

Around the beginning of June, pressure patterns begin to change in line with the northward movement of the **Inter Tropical Convergence Zone** (ITCZ). A deep low pressure centred at the heart of the ITCZ advances northwards in a series of steps. The rains are not uniformly distributed over the regions of India, but advance from the SW in two distinct streams. Areas such as southwestern India and parts of Bangladesh receive the monsoon at the beginning of June as they lie in the direct path of the SW winds.

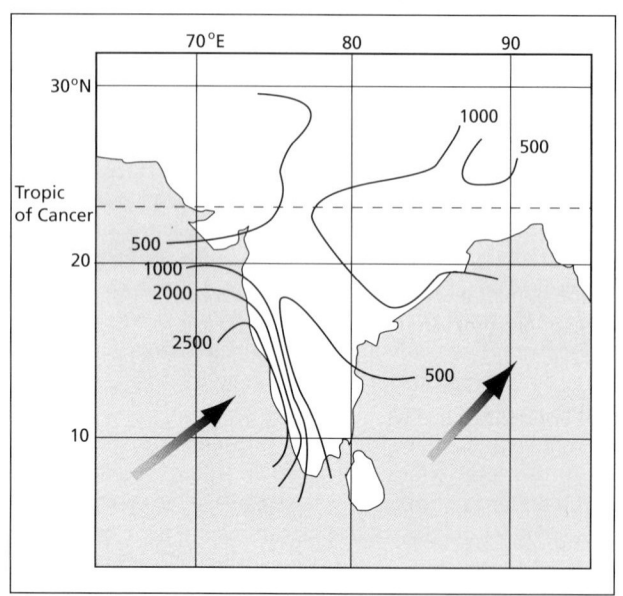

Resource 11: Average monsoonal precipitation (June-Sept) (Based on information from *Contemporary Climatology* Robinson and Henderson 1999)

Rains in the east and central India do not arrive until mid-June while in areas such Kashmir and Gurjurat it could be early July before the rains arrive, if at all!

Over NW India the intense heating produces a thermal low (**Resource 11**), while in the relatively cooler temperatures of the Indian Ocean a high-pressure system is produced. This difference in pressure creates a **pressure gradient** and winds blow from high pressure to low pressure. Winds are drawn in from the equator and change direction (becoming south-westerly) due to the influence of the Coriolis force which is one factor in producing the SW monsoon.

The winds crossing the Indian Ocean and the Bay of Bengal become **moisture laden** and will be capable of producing large amounts of rainfall when they reach the continental land mass. To sustain this pattern of **convergence** of winds at the surface, there is divergence of air aloft (above 10 km) in the form of a jet stream. This high-level air stream aloft enables air to be removed from the region and permits a continued inflow at the surface (**Resource 12**).

Finally, in order to explain the sudden onset of the monsoon we must look to the effect of the Himalayas. This vast mountain barrier acts like a block on the seasonal shift of the pressure belts towards the poles. As June approaches, pressure grows on the wind belts to move northwards, causing the jet stream to shift north too. However, the westerly jet that flows in this region is trapped on the equatorward side of the mountain, maintaining a high-pressure system well into spring. Then, almost overnight, the jet stream suddenly shifts north of the Himalayas allowing low pressure to build rapidly and enabling the wet monsoon rains to be drawn into this region. The need to forecast the onset of the monsoon has become a major preoccupation of the Indian meteorological services, due to its significance for so many millions of farmers and for the economy as a whole.

North East Monsoon (winter monsoon)

During the months of November to May, as the **ITCZ** moves southwards over the equator (see **Resource 11**), intense low temperatures in central Asia produce a high pressure. The very low temperatures (as low as -34^0C in central Asia) are a consequence of the impact of continentality experienced by the region in winter. At the same time, a deep low-pressure system builds over central Australia creating a different **pressure gradient** from that which evolved in the summer. This time a reversal of winds occurs, drawing NE winds out from the high-pressure regions of central Asia. After crossing the subcontinent, these winds begin to curve under the influence of the **Coriolis force** to become north-westerlies as they cross the equator. The fact that the source of air is a continental land mass ensures the air is **dry** and produces little precipitation in the regions through which it flows. Most states in India experience this dry season between November and May except for those in the south-east, which receive rainfall evaporated from the Bay of Bengal. These dry air streams produce clear skies and hot weather.

As the pressure belts move south, the **westerly jet** splits into two across the Himalayas, with an air flow to the north and one to the south of the mountain barrier. This permits the normal shift of pressure belts to the south.

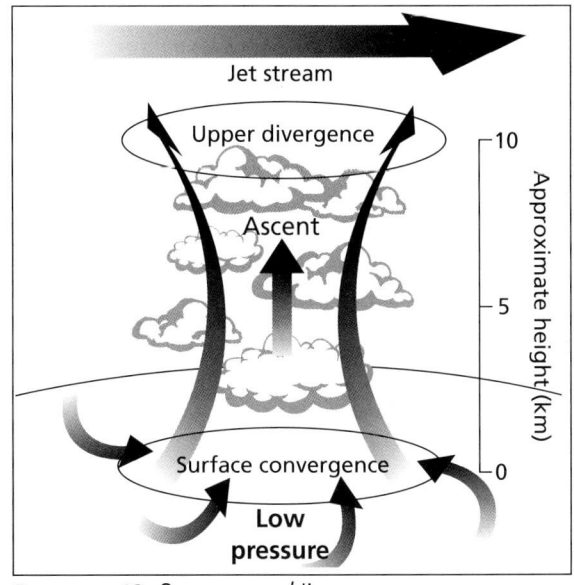

Resource 12: Convergence/divergence
(Based on information from *Weather Systems* Musk 1988)

The problem of late arrival of the rains – Indian subcontinent

The variability of the monsoon rains is a major hazard with devastating implications for the people and the economy of this nation. The time of arrival of the monsoon rains determines the length of the rainy season and that in turn determines the levels of crop yield. With a population of almost one billion and 66% dependent on agriculture, the arrival of the monsoon in India is vital in supplying food to the population. The two hazards are drought and floods. If the rains arrive late and the season is shortened, drought may be the outcome. In drought years the crops, in particular rice, will fail and the resulting famine will not only damage the economy but also lead to many deaths throughout the country. If the rains come early and the season is lengthened, flooding becomes the hazard in low-lying areas such as the lower Ganges Valley.

As one travels north and westward in India, the monsoon duration period becomes reduced. For example, Bangalore has 5 months of the monsoon rains and Bombay has 4 months, while places such as Bikaner in the NW of India have only 2 months. The annual variation in rainfall at Bombay can range from as low as 600 mm (1918) to more than 3400mm (1954). Scientists have discovered that the intensity of droughts over India is closely associated with El Niño episodes, while flooding is more difficult to predict. Forecasting droughts in particular has become vitally important in India today.

Social and economic impacts

Areas of central and northern India are most vulnerable to the drought hazard because here the monsoon rains are lightest and least reliable. During the 1980s the rains 'failed' in a number of years, especially in 1987 and 1989. The northern regions were very badly affected, not only because the crops failed but also because so many were subsistence farmers whose livelihood depended on the rice crop yield. Land reform schemes in areas such as the Punjab have enabled many farmers to invest in irrigation systems, which help combat the threat of drought in dry years.

Sometimes the rains arrive prematurely and then cease for a while before recommencing. This situation can have serious consequences for the crop yields. As soon as the rains begin many farmers will plant their rice crops to avail of every drop of moisture. However, with cessation of the rains, these crops, which require lots of moisture for survival, will shrivel up and die. Given the lack of financial resources available to many farmers, the ability to purchase new seed and replant later is not a viable option. The outcome is financial ruin for many of the poorer farmers and a possible food crisis for that region, if the majority of farmers have all planted at the same time. With no crop and no savings many farmers and farm labour are forced off the land.

Flooding can be another hazard, especially in the lower Ganges valley because late monsoon clouds, instead of moving towards the Himalayas as usual, may dump as much as 1700 mm of rains in the area below Farakka Barrage in West Bengal in little more than ten days. Such events can submerge areas for weeks and cause the loss of life and great disruption to the infrastructure. This is a common problem for countries such as Bangladesh.

The problem of river flooding and storm surges – Indian subcontinent

Introduction

Bangladesh is located in south-east Asia in the delta of the rivers Ganges, Bramaputra (Jamuna), Meghna and numerous other small rivers that drain into the Bay of Bengal. It has a growing population of 123 million people. Approximately 80% of the country lies in the delta and is less than six metres above sea level. Almost all of Bangladesh's rivers originate outside the country, eg the rivers Ganges and Brahmaputra. Seventy-five percent of the annual rainfall occurs in the summer monsoon season between June and September. Peak discharges of these three rivers are huge – approximately 100,000 cumsecs. All the rivers carry a large amount of sediments, which are annually deposited in the delta region creating islands and sandbanks along the delta.

Types of floods

In fact, floods created Bangladesh. Over millions of years, the Himalayas have been eroded by numerous rivers flowing into the delta region dumping silt there. Bangladesh suffers flooding on an annual basis. This can be regarded as normal and usually covers between a third and a half of the country during the monsoon season. It is helpful to distinguish between 'normal' (*barsha*) floods to which agricultural practices are well adapted and the 'damaging' (*bonna*) floods when water rises earlier, higher and faster than expected.

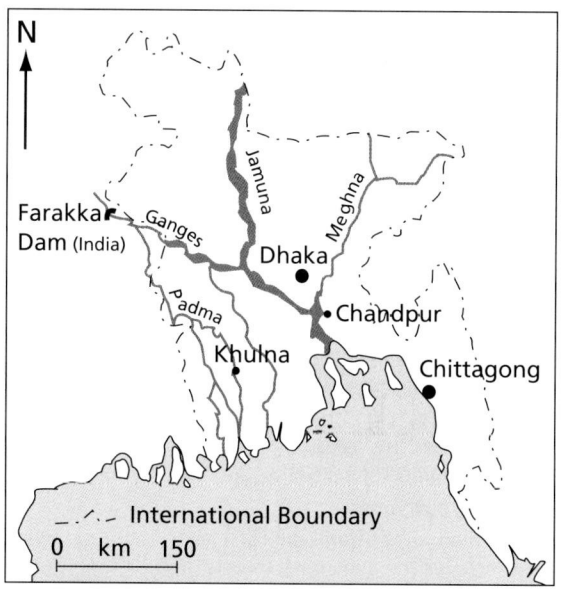

Resource 13: Map of Bangladesh, Farakka Dam

The intensity of the summer monsoon rains (June to September) produces **rainwater flooding** on an annual basis and affects between a third and a half of the country overall. These could be regarded as normal floods and happen because of the amount of the rain and the flatness of the countryside. The extent of any damage to crops depends on the time when the floods occur.

Flash flooding, resulting from exceptionally heavy rainfall over the neighbouring mountains, tends to affect areas in the north and east, where rivers from India enter the country. Some researchers believe dam building in India (Farakka Dam) and increased urbanisation have contributed to damage to crops and extra sedimentation.

River floods are fairly common and are produced by snowmelt in the Himalayas leading to peak discharges in all the main rivers draining into the delta region (**Resource 13**). The river channels in many cases cannot cope and when this surge of water reaches the floodplain, naturally flooding occurs. If peak discharges combine with the monsoon rains, then severe flooding usually occurs in Bangladesh, eg August–September 1998.

Storm surges are a constant hazard in the delta region and islands of Bangladesh. Outside the wet season the country receives heavy rainfall due to depressions tracking in from the Bay of Bengal. Every so often, tropical cyclones (deep low-pressure systems) form and move inland. Storm surges are caused by such systems, which suck up the sea by up to four metres and move it inland. Incoming surges last for a few hours and only affects an area of between 4-8 kms inshore, but their impact can be devastating in terms of loss of life and damage to property. In 1991 it was estimated that 150,000 people died on the coastal islands as a result of such a storm.

Social and economic impact of flooding

Beneficial outcomes:
- Economically, river flooding replenishes the fertility of the land which enables extra crops to be grown, increasing the income for farmers. The increased fertility can mainly be attributed to the nitrogen-fixing blue-green algae which live in the water and feed on the decomposing waterlogged plants. Over time silt also increases soil fertility. It has been found that crop yields increase significantly the year after a flood.
- Flooding replenishes the groundwater supplies and provides long-term water stores for domestic purposes in many regions. This has a positive impact on the health and the livelihood of people in Bangladesh.
- The floods provide a resource – fish that are farmed and used as a food source to supplement a diet which consists mainly of rice.
- The flooding also flushes out pollution and pathogens on the land areas and within the rivers.
- Some rivers are used as dumping grounds for industrial waste products, as bathing places and as navigation routes for ships and transportation of goods.

Negative outcomes
- The high percentage of flat land and poor infrastructure in Bangladesh makes evacuation to high land more difficult. Many people lose their lives in the annual floods and this is increased to thousands when a tidal surge is the cause. In 1998 a combination of different types of flooding hit Bangladesh at the same time and created devastating floods covering 57% of the country. Approximately 1000 people died and thousands were made homeless. Nevertheless, in comparison to the death toll from storm surges this might be regarded as small. In 1970 a storm surge hit the coastal regions killing 300,000 people.
- When flooding is severe property and land are inundated and crops destroyed by the rising floodwaters. Staple food crops such as rice are lost and so food shortages, famine and starvation for thousands follow a flood event such as that in 1998.
- Floods can cut off regions for days as bridges and roads become impassable or are

destroyed completely. In 1998 almost half of the capital city, Dhaka, was submerged for up to two months. Even after the waters had receded, villages were buried in sand and silt and required lots of resources to revitalise them again.
- Secondary impacts include diseases associated with the lack of available clean water. In the aftermath of a storm surge the number of unburied corpses presents a major health risk. Many of the weakest (the very young and the very old) in the society fall ill and with little medical help available, succumb to cholera or typhoid.
- Economically, floods use up a nation's wealth reserves. In many cases famine forces the government to purchase expensive imports of rice to ensure people's survival until the next crop. Being a poor country and in debt, Bangladesh's financial reserves are low. It was estimated that the total cost of the 1998 floods was almost $1 billion.

Human responses to the flood threat in Bangladesh

The Bangladeshi government has always tried to address the immediate impacts by distributing food and providing regional relief in the form of clean water and sanitation services. However, the responses have been inadequate due to the scale of the problem, the huge numbers affected (*Bangladesh has a population density of 803 people per km^2 while in Ireland it is just 50 per km^2*), and the lack of money and resources. Such solutions have been short-term and partial. Longer-term options have included a structural approach, a 'soft' engineering approach, and some believe that a social/political approach is the way forward.

The structural approach focused on **flood prevention** measures, **drainage improvements** and **flood proofing** projects. After the disastrous floods of 1988, the World Bank prepared an **Action Plan** (July 1989) to alleviate the effects of flooding in Bangladesh. The main flood prevention strategy embodied in this Plan was the construction of 3,500 km of embankments to control the level of flooding and the designation of compartments for floodwater storage. Another aim was to strengthen and raise the sea defences along the coast to reduce the damage caused by storm surges. In addition, the Plan included the building of seven large dams, 12-15 floodwater storage basins to hold floodwaters diverted from the main rivers and the replanting of forested areas in the Himalayas. Dhaka and other Bangladeshi cities would be protected up to a flood frequency event of between 500 and 1000 years. The economic cost of this Action Plan has been borne by donor countries but the work is being carried out by agencies of the Bangladeshi government.

Drainage improvement responses include dredging riverbeds to keep the channels open and sluice gates to divert the flow of water away from high value areas. Although costly techniques, both are relatively effective. Maintenance has proved to be a real problem however.

Another option was flood proofing, entailing projects such as dam-building and irrigation schemes to limit the need for floodwaters for economic activities. This however, has had mixed outcomes. Studies from the Chandpur Irrigation Project show that whilst the average income and yields improved within the project area, the benefits were not evenly distributed. The large landowners tended to benefit most, while the poorer tenant farmers and landless labourers in the area gained little from the investment. Some see this strategy as too costly and too long-term for an economically poor country like Bangladesh.

In the coastal region and on the islands such as Bhola, Sandwip polders and embankments have been constructed. But polders constructed in these coastal areas to keep out the intrusion of saline water from the Bay depend upon an intricate network of sluice gates. Many of these have become inoperative, requiring urgent repair. The experts warn that cutting of the polders to release floodwaters at any point might make these areas vulnerable in the future to tidal surges from the Bay of Bengal.

FLOODING

Benefits of embankments	Benefits of flooding
• Reduces deaths • Reduction in damage to property • Reduction in the spread of disease • Industries protected and employment preserved • Infrastructure protected • Reduction in land lost by erosion	• Silt enriches the soil • Floods maintain lake areas where fish (protein) breed • Soils that have been enriched by floods produce increased yields in the dry season • Floodwaters recharge the groundwater supplies • Floodwaters drain easily back into the channel after flood event

There are those who believe that such 'hard' (structural) responses are not the way forward and argue for a 'soft' engineering approach, that would be more in keeping with the appropriate technology required by a poor country like Bangladesh. Building higher embankments is the wrong strategy because of the difficulties they pose. These include:

- the retention of floodwater which prolongs the length of flood times. Subsequently, the stagnant water produced will pose health problems for the local population. Evidence of this could be seen in coastal areas after the storm surge of 1991, when the return of floodwaters was delayed behind roads and embankments. Freshwater supplies were contaminated with salt from the seawater and disease became widespread.
- sudden breaches of these high embankments which are likely to cause more damage and deposit deep layers of infertile sand, reducing soil fertility.
- compartmentalisation projects which reduce the flushing effect of the floodwaters.

Those advocating a less technological, more ecocentric approach call for better weather flood forecasting services, local warning schemes and improved flood shelters. They believe in more localised flood management involving the local community. New ponds could be built for fishing and water storage, and low embankments would check flooding only during the early growth stages of the rice crop. Ecosystems would be protected and sustained and in some areas natural floodplain processes would be allowed to develop. Improved emergency services are seen as essential when unusually high and rapid flooding occur. All these are viewed as a cheaper alternative to the costly, 'top-down' strategies proposed by the Action Plan.

Another school of thought supports a more radical, deep-rooted response which takes account of the social/political environment and addresses the causes of why people are vulnerable in the first place. The argument asserts that a disaster is composed of the natural event and the human response to it. In Bangladesh the floods have such social/economic impacts because of the way people react to this natural hazard. Flooding is a disaster in Bangladesh not simply because of the rainfall levels, but because the people are trapped by poverty and have few resources to cope when disaster strikes. Many make their living as landless labourers and are vulnerable to flood-induced unemployment. Much of the housing consists of illegal squats in low demand flood-prone areas. Because they have so little, they are more vulnerable. This approach advocates greater cooperative organisation of the poor to gain resources, better access to land for all, compensation for animals and other assets lost due to the flood hazard, and better healthcare for the majority of the population. Vulnerability is the key issue that requires answers.

In conclusion, the flood hazard in Bangladesh needs to be seen as a complex issue and not merely as a climatic-hydrological problem. The Flood Action Plan has cost many billions and has helped alleviate some of the problems. However, a structural approach cannot be the complete solution as the experience with the Mississippi river (USA) showed. For some of the projects to be successful, international cooperation with neighbouring states is essential. Given the animosity between Bangladesh and some of its neighbours, this seems an unrealistic goal in the foreseeable future. In addition, if global warming becomes a reality in the near future then such large investments may be a waste of resources in a low-lying country like Bangladesh.

A more ecocentric approach with local involvement must also be interwoven into any solution. But again, the issue of vulnerability cannot be ignored. Some view the flooding issue as a symptom of a much deeper malaise, which will prove a lot harder to solve. Frampton et al (1996) state:

> *The real hazard in Bangladesh is not the occasional floods but the daily struggle with malnutrition and poverty.*

River floods kill on average 500 people per year in Bangladesh, but how many die of starvation and disease because they have neither the resources nor the political power to change things? Such radical questions demand serious consideration.

References and further reading

Alexander, D (1993) *Natural Disasters* University College London
Bolt, B (1988) *Earthquakes* W H Freeman and Co
Bowen, A/Pallister, J (2000) *AS Level Geography* Heinemann
Bryant, E (1991) *Natural Hazards* Cambridge University Press
Buckley, R (1993) *Earthquakes and Volcanoes: Living with the Restless World* Understanding Global Issues European Schoolbooks
Decker, R & B (1991) *Mountains of Fire* Cambridge University Press
Digby, B (2000) *Changing Environments* Heinemann
Fielding-Smith, A and Smith, R (April 1996) *Earthquakes – An Update* Geofile Stanley Thornes
Fifield, R (editor) (1985) *The Making of the Earth* Basil Blackwell
Foskett, N & R (1987) *Earthquakes and Volcanic Eruptions- Prediction and Response* Geofile Mary Glasgow
Frampton, S, Chaffey, J, McNaughton, A, Hardwick, J (1996) *Natural Hazards* Hodder & Stoughton
Francis, P (1993) *Volcanoes: A Planetary Perspective* Clarendon Press
Guiness, P and Nagle, G (1999) *Advance Geography* Hodder & Stoughton
Hallam, A (1973) *A Revolution in the Earth Sciences* Oxford University Press
Hallam, J, Handsley, M and Hilton, K (1983) *Earthshaping – The Lively Earth* University Tutorial Press
Jeffcoat, A and Robinson, B (Sept 1997) *Volcanic Activity-Causes and Consequences* Geo Factsheet
Nagle, G (1998) *Hazards* Nelson
Newstead, L (April 2001) *Montserrat Volcanic Eruptions 1995-98* Geofile 401 Nelson Thornes
Palutikof, J, Subak, S and Agnew, M (1997) *Economic Impacts of the Hot Summer of 1995* Dept of the Environment
Park, C C (1991) *Environmental Hazards (2nd Ed)* Macmillan
Ollier, C (1989) *Volcanoes* Basil Blackwell
Robinson, A (1993) *Earthshock* Thames and Hudson
Ross, S (1998) *Natural Hazards* Stanley Thornes
Scarth, A (1997) *Savage Earth* Harper Collins
Smith, K (1992) *Environmental Hazards* Routledge
Smith, R and Fielding-Smith, A (1998) *Volcanoes – Recent Research and Case Studies* Geofile Stanley Thornes
Understanding Global Issues European Schoolbooks
Video 'Pinatubo' Horizon
Warburton, P (1995) *Atmospheric Processes and Human Influence* Collins Educational
Whittow, J (1980) *Disasters. The Anatomy of Environmental Hazards* Pelican
Witherick, M (1999) *Environment and People* Stanley Thornes
Wright, L (1993) *Environmental Systems and Human Impact* Cambridge
There are numerous websites dedicated to these topics including details of the Kobe, Pinatubo and Gujarat events, eg http://www.gujaratplus.com/news/archives/arc83.html

Articles
1998 (Apr) *Geo Factsheet* (45) – Flood Management in Bangladesh
1988 (Jan) *Geofile* 99 Bangladesh – E Wallis
1999 (April) *Geofile* 355 Anticyclonic weather in the UK – J Pallister
2000 (Jan) *Geo Factsheet* (85) – Managing the River Ganges – Water Quality